CHERRY LAKE PRESS

Published in the United States of America by Cherry Lake Publishing Group
Ann Arbor, Michigan
www.cherrylakepublishing.com

Reading Adviser: Beth Walker Gambro, MS, Ed., Reading Consultant, Yorkville, IL
Book Design: Jennifer Wahi
Illustrator: Jeff Bane

Photo Credits: © Yuriy2012/Shuttershock.com 2, 3, 24; © Antoniya Kadiyska/Shuttershock.com, 5; © dugdax/Shuttershock.com, 7; © Olga_Olechka/Shuttershock.com, 9; © WildMedia/Shuttershock.com, 11, 23; © BEJITA/Shuttershock.com, 13; © Leena Azzam/Shuttershock.com, 15; © Pawel Brud/Shuttershock.com, 17; © Charles Plant/Shuttershock.com, 19; © Sergey Uryadnikov/Shuttershock.com, 21; Cover, 10, 16, 22, Jeff Bane

Copyright © 2023 by Cherry Lake Publishing Group
All rights reserved. No part of this book may be reproduced or utilized in any form or by any means without written permission from the publisher.

Cherry Lake Press is an imprint of Cherry Lake Publishing Group.

Library of Congress Cataloging-in-Publication Data

Names: Gray, Susan Heinrichs, author. | Bane, Jeff, 1957- illustrator.
Title: Forest / Susan H. Gray ; illustrated by Jeff Bane.
Description: Ann Arbor, Michigan : Cherry Lake Publishing, 2022. | Series: My guide to earth's habitats | Audience: Grades K-1
Identifiers: LCCN 2022005367 | ISBN 9781668908938 (hardcover) | ISBN 9781668910535 (paperback) | ISBN 9781668912126 (ebook) | ISBN 9781668913710 (pdf)
Subjects: LCSH: Forest ecology--Juvenile literature.
Classification: LCC QH541.5.F6 G725 2022 | DDC 577.3--dc23/eng/20220214
LC record available at https://lccn.loc.gov/2022005367

Printed in the United States of America
Corporate Graphics

table of contents

Trees!..............................4

Glossary 24

Index 24

About the author: Susan H. Gray has a master's degree in zoology. She loves writing science books, especially about animals. Susan lives in Arkansas with her husband, Michael. They think forests are very peaceful.

About the illustrator: Jeff Bane and his two business partners own a studio along the American River in Folsom, California, home of the 1849 Gold Rush. When Jeff's not sketching or illustrating for clients, he's either swimming or kayaking in the river to relax.

Trees!

A forest is full of trees. Some trees have leaves. Others have **needles**.

Often, the trees are crowded.
Sunlight cannot get through.
The forest floor is shady.

Ferns love the shade. **Violets** like it, too.

Many animals are at home in the forest. They find food and water. They also find **shelter**.

Beetles live on the forest floor. **Termites** hide in stumps. Butterflies fly through the trees.

Cardinals make nests here.
They eat insects and seeds.
The forest has all they need.

Woodpeckers are here, too. They peck holes in trees. Ants and termites live in the tree bark. The birds gobble them up.

Squirrels eat **acorns** and nuts. They build nests in trees. Nests are made of leaves and twigs.

Big animals also live here.
Bears look for food. Deer and
foxes do, too.

The forest is a great place. Plants and animals are everywhere!

glossary & index

glossary

acorns (AY-kornz) the nuts of oak trees

cardinals (KAR-duh-nulz) red birds that live in parts of North America, Mexico, and Central America

needles (NEE-duhlz) the thin, stiff leaves of pine or fir trees

shelter (SHEL-tuhr) a safe, protected place

termites (TUHR-myts) antlike insects that feed on wood

violets (VY-uh-luhts) short, leafy plants that often have blue or purple flowers

woodpeckers (WUHD-peh-kuhrz) insect-eating birds with hard beaks that can pierce wood

index

animals 10, 20, 22

floor, 6, 12

leaves, 4, 18

nests, 14, 18

shade, 6, 8

termites, 12, 16

trees, 4, 6, 12, 16, 18